AF575804

Julie K. Lundgren

A Crabtree Seedlings Book

Crabtree Publishing
crabtreebooks.com

TABLE OF CONTENTS

INSECT OR SPIDER?

Spiders have eight legs and two body sections. Spiders are **arachnids**, not insects. An insect has six legs and three body sections.

SPIDER BODY SECTIONS
abdomen
cephalothorax

Many spiders have eight eyes. They must have good eyesight, right? Not usually. They can sense light and dark.

CREEPY BUT COOL CLOSE-UP

WHAT GREAT EYES, AND EYES, AND EYES!
jumping spider

Spiders can be tiny, huge, smooth, hairy, slim, or fat.

The tiny jumping spider can fit on a fingernail.

The Goliath bird-eating tarantula, the world's largest spider, barely fits on a dinner plate.

Some spiders' hairs increase their sense of touch. Others sense sound waves, a bit like little ears.

Hairs cover the Green bottle blue tarantula.

Spiders have different colors and patterns. Some blend in. We call this **camouflage**.

Beach wolf spiders hide in plain sight.

Smooth flower crab spiders match flowers.

DEADLY HUNTERS

All spiders are **carnivores**. Many hunt insects, while others eat mice, birds, or other spiders.

banded argiope spider

Some spiders use silk from their spinnerets to spin webs to catch their prey. A spider's spinnerets can make different types and thicknesses of silk.

tarantula

Different kinds of spiders make different kinds of webs.

Messy webs: *strands in no order*

Sheet webs: *a layer of silk, often on top of grass or bushes*

Funnel webs: *like sheet webs, except with a central tunnel where the spider hides*

Orb webs: *sticky circles inside of circles hanging between twigs, buildings, or other structures*

Spiders that don't use webs **ambush** their prey.

Fishing spiders launch underwater attacks on passing insects or small fish.

Most spiders don't eat their prey whole. Some use their fangs to inject **venom**. Venom turns the insides of prey into liquid, so the spider can drink it like soup.

LIFE CYCLE

Spiders wrap their eggs in a silk egg sac.

Some spiders may lay only one or two eggs.

Some may lay over two thousand!

When they hatch, the spiderlings look like tiny adults.

Growing spiders **molt** several times. Their hard **exoskeletons** slip off to reveal new, larger ones.

Most spiders live only one or two years. Large spiders like tarantulas can live more than twenty years.

Many people have these creepy but cool predators as pets. Would you?

Mexican Red Knee Tarantula

LABEL THE SPIDER

Match each word below with the correct box:

abdomen
cephalothorax
legs
spinnerets

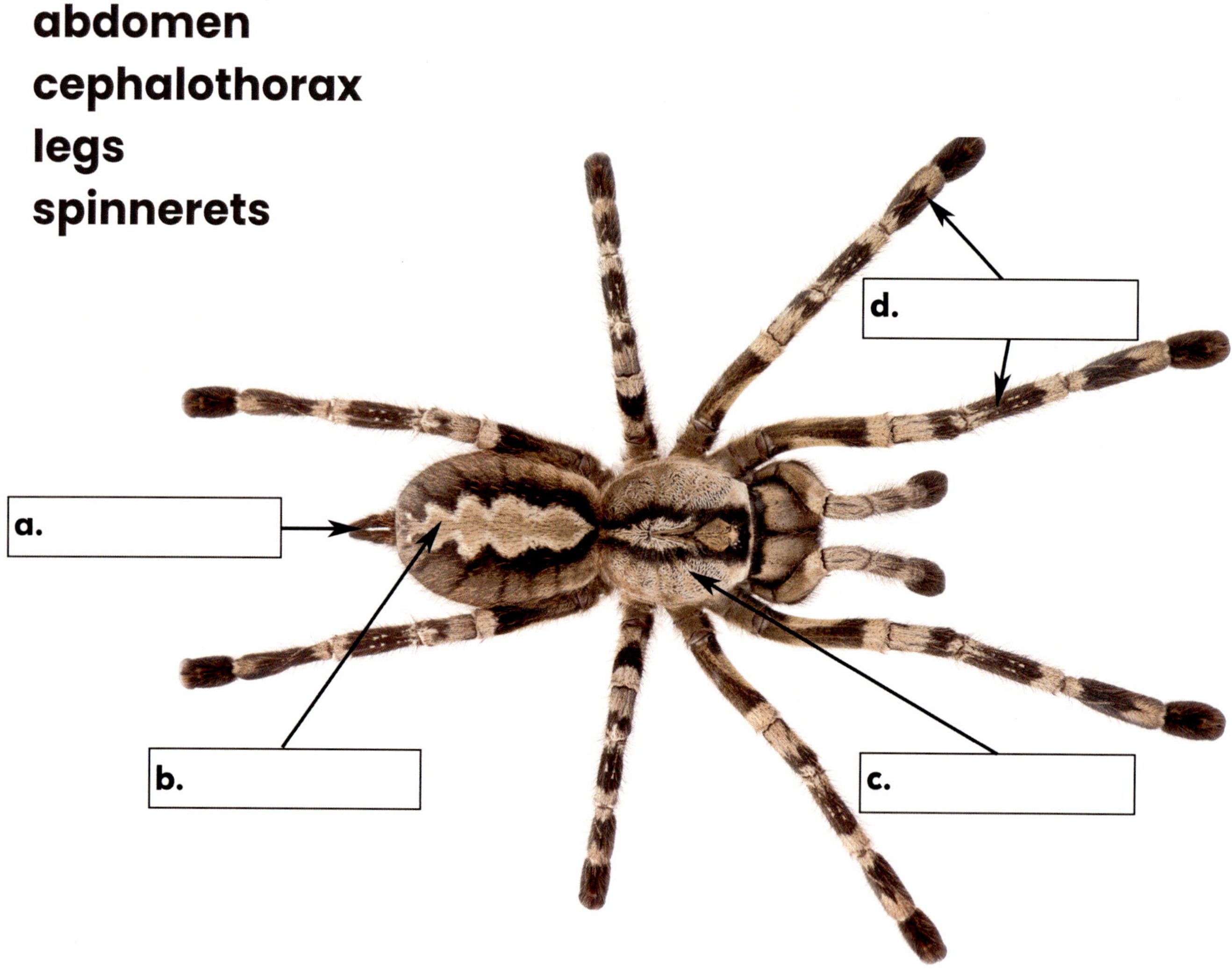

Answers: **a.** spinnerets, **b.** abdomen, **c.** cephalothorax, **d.** legs

ambush (AM-bush): To attack by surprise, like a sneaky ninja

arachnids (uh-RAK-nidz): Animals with eight legs, two body sections, and no wings or antennae

camouflage (KAM-uh-flahzh): Animal colors and patterns that blend with the surroundings to help animals stay hidden

carnivores (KAR-nuh-vorz): Animals that catch and eat other animals

exoskeletons (ex-oh-SKEL-uh-tunz): The hard, outer cases that support the body in animals without bones

molt (MOHLT): Shed skin or an outer covering, so the animal can grow

venom (VEN-um): Poison used to stun or kill prey, delivered by a bite

INDEX

School-to-Home Support for Caregivers and Teachers

This book helps children grow by letting them practice reading. Here are a few guiding questions to help the reader build his or her comprehension skills. Possible answers appear here in red.

Before Reading

- **What do I think this book is about?** *I think this book is about many different kinds of spiders. I think this book is about how spiders make webs.*
- **What do I want to learn about this topic?** *I want to learn how a spider makes a web. I want to learn if there are any poisonous spiders that can kill humans.*

During Reading

- **I wonder why...** *I wonder why spiders have eight eyes and eight legs. I wonder why some spiders' hairs can sense sound waves like little ears.*
- **What have I learned so far?** *I have learned that all spiders are carnivores. I have learned that most spiders inject venom into their prey that turns the insides of the prey into liquid that the spider drinks.*

After Reading

- **What details did I learn about this topic?** *I have learned that most spiders live only one or two years. I have learned that large spiders like tarantulas can live more than twenty years.*
- **Read the book again and look for the glossary words.** *I see the word **arachnids** on page 4, and the word **molt** on page 20. The other glossary words are found on page 23.*

Crabtree Publishing

crabtreebooks.com 800-387-7650

Print book version produced jointly with Blue Door Education in 2022

Content produced and published by Blue Door Publishing LLC dba Blue Door Education, Melbourne Beach FL USA. Copyright Blue Door Publishing LLC. All rights reserved. No part of this book may be reproduced or utilized in any form or by any means, electronic or mechanical including photocopying, recording, or by any information storage and retrieval system without permission in writing from the publisher.

Written by Julie K. Lundgren
Production manager: Candice Campbell

Hardcover 978-1-4271-6170-3
Paperback 978-1-4271-6182-6

Printed in the U.S.A./CP052026

Library and Archives Canada Cataloguing in Publication
Title: Spiders / Julie K. Lundgren.
Names: Lundgren, Julie K., author.
Description: Series statement: Creepy but cool | "A Crabtree seedlings book". | Includes index. | Previously published in electronic format by Blue Door Publishing FL in 2015.
Identifiers: Canadiana (print) 20210201827 | Canadiana (ebook) 20210201835 | ISBN 9781427161703 (hardcover) | ISBN 9781427161826 (softcover) | ISBN 9781427161949 (HTML) | ISBN 9781427162069 (EPUB) | ISBN 9781427162182 (read-along ebook)
Subjects: LCSH: Spiders—Juvenile literature.
Classification: LCC QL458.4 .L86 2022 | DDC j595.4/4—dc23

Published in Canada
Crabtree Publishing
616 Welland Avenue
St. Catharines, Ontario
L2M 5V6

Published in the United States
Crabtree Publishing
347 Fifth Avenue
Suite 1402-145
New York, NY 10016

Photo credits: Cover ©shutterstock.com/Pablo Hidalgo - Fotos593, page 2 © istockphoto/ ConstantinCornel, page 4-5 ©shutterstock.com/James L. Davidson, video ©GlobalStock, page 6-7 ©shutterstock.com/skynetphoto. page 8 ©shutterstock.com/89studio, page 9 dinner plate ©shutterstock.com/ILYA AKINSHIN, fork © studioVin, Goliath spider © Snakecollector. page 10 © Cathy Keifer, page 11 © Amir Ridhwan, crab spider ©shutterstock.com/Lidara. page 12-13 ©shutterstock.com/Cathy Keifer. page 14 ©shutterstock.com/Eric Isselee. page 15 messy web © J.Gatherum, sheet web ©Ilya Sirota, funnel web © PHOTO FUN, orb web © Mark Carthy, Page 16 © Brian Lasenby. Page 17 fangs © D. Kucharski K. Kucharska, Page 18 wrapping eggs in silk © Cathy Keifer. Page 19 spiderlings © Dr. Morley Read, Page 20 spider molting ©shutterstock.com/noppharat; Page 21 ©shutterstock.com/cowboy54. Page 22 © Eric Isselee

Library of Congress Cataloging-in-Publication Data
Names: Lundgren, Julie K., author.
Title: Spiders / Julie K. Lundgren.
Description: New York : Crabtree Publishing, 2022. | Series: Creepy but cool - a Crabtree seedlings book | Includes index.
Identifiers: LCCN 2021018448 (print) | LCCN 2021018449 (ebook) | ISBN 9781427161703 (hardcover) | ISBN 9781427161826 (paperback) | ISBN 9781427161949 (ebook) | ISBN 9781427162069 (epub) | ISBN 9781427162182
Subjects: LCSH: Spiders--Juvenile literature.
Classification: LCC QL458.4 .L86 2022 (print) | LCC QL458.4 (ebook) | DDC 595.4/4--dc23
LC record available at https://lccn.loc.gov/2021018448
LC ebook record available at https://lccn.loc.gov/2021018449